The Sound Files

Discovery Channel School
Science Collections

Discovery Communications, Inc., produces high-quality television programming,
interactive media, books, films, and consumer products. Discovery Networks, a division of Discovery
Communications, Inc., operates and manages Discovery Channel, TLC, Animal Planet, Discovery Health Channel, and Travel Channel.

Writers: Jackie Ball, Justine Ciovacco, Bill Doyle, Ruth Greenstein, John-Ryan Hevron, Scott Ingram, Monique Peterson, Rachel Roswal. **Editor:** Justine Ciovacco.
Photographs: cover, guitar, ©Patrick Cocklin/Tony Stone Images; pp. 2, 4–5, Dizzy Gillespie, ©Corbis-Bettmann/UPI; pp. 3 & 14, ear trumpet, ©CORBIS; pp. 10–11, piano
with hands, ©DigitalVision/PictureQuest; p. 12, Flea, ©Capital/Longhorst/CORBIS; Beethoven, ©Bettmann/CORBIS; p. 13, rock band, ©Tony Stone Images/Christopher
Bissell; piano, ©ARTVILLE; guitar, ©2000 Gibson Musical Instruments; p. 14, 1900 hearing aid, ©Bettmann/CORBIS; p. 15, 1940 hearing aid, ©UPI/CORBIS-Bettmann; 1999
hearing aid, ©AFP/CORBIS; p. 16, courtesy Beverly Biderman; p. 19, Bell & Watson, ©PR/Science Photo Library; p. 19, phone, ©Library of Congress; p. 22, killer whale,
©Painet Inc.; basketball player, ©Comstock; p. 23, cricket, ©Painet Inc.; glassfish, ©Mark Smith/Photo Researchers; snake charmer, ©Painet Inc; p. 26, wine glass;
©Steven Dalton/Photo Researchers, Inc.; ultra sound, ©SuperStock, Inc.; pricked finger, ©Photo Researchers, Inc.; p. 29, J. Cochran & J. Auriol, ©Bettmann/CORBIS;
p. 31, anechoic chamber courtesy Macintosh. **Illustrations:** p. 10, inner ear, Steve Stankiewicz; p. 17, implant, Steve Stankiewicz; p. 20, map symbols, Rita Lascaro;
pp. 24–25, Dave Winter. **Acknowledgments:** pp. 16–17, WIRED FOR SOUND. Beverly Biderman. Trifolium Books, Inc., 1998. Reprinted with permission; pp. 26–27, Jackie
Cochran: AN AUTOBIOGRAPHY. Jacqueline Cochran and Maryann Bucknum Brinley. Bantam Books, 1987. Reprinted with permission; pp. 26–27, I LIVE TO FLY.
Translated from the French by Pamela Swinglehurst. E. P. Dutton & Co., Inc., 1970. Reprinted with permission.

CONTENTS

Vibes

Listen. What do you hear? Even if no one's around and your answer is "silence," you can be sure that sound waves are bouncing around you. You're breathing, right? You'll have to turn the page, close the book, or put it down at some point, right?

Sound waves are all around us in the form of vibrations. A partnership between the brain, ear, and structures in between helps us to interpret these vibrations as sounds. Yet there is also a whole world of sounds that we *can't* even hear. Other animals, however, may tune in on them. That's why a dog may pick up its head and look around, even though you don't see or hear anything.

In VIBES, Discovery Channel takes you on a journey through the hearing process and beyond. Learn how sound is produced and picked up by ears, what it's like to be unable to hear, how the amazing energy of sound waves can heal or harm, and more.

The Sound Files

Wake-Up Call 6
Q & A We've found a rooster with something to crow about. He's proud of the noise . . . er, um, sound . . . he creates. So proud that he's eager to explain how he produces the sound and how you hear it.

The Wave of the Present 8
Almanac Take a look at a sound wave and find out what makes it change its tune—and appearance. Plus, find out what sounds humans can commonly produce and safely hear.

Playing Along 10
Virtual Voyage Take a bumpy journey through the ear canal on the "wings" of a sound wave.

Music, Sweet Music 12
Scrapbook Unfortunately, as some musicians have found out the hard way, sound's vibrations can shake them of their most precious gifts: hearing and creating music. See how they have learned to cope. Plus, see how instruments use vibrations.

Catching the Vibe 14
Timeline Follow the history of hearing helpers for the sound impaired.

Wired For Sound 16
Eyewitness Account Beverly Biderman has lived without hearing all of her life. After many years of trying to cope in a world full of sounds, she received a cochlear implant that allowed her to hear. She explains what it's like to enjoy sound for the first time.

Can a horn help you hear? See page 14.

Final Project

Vibes

Check out jazz trumpeter Dizzy Gillespie. You can be sure he was hard at work when this picture was taken. His job? Creating vibrations.

Sound is the result of a vibration. It travels from its source—in this case, Dizzy's vibrating lips—in invisible waves. Trumpet players use their breath and lips to create the vibrations that produce sound. A musician's lip pressure on the mouthpiece can make the air inside the instrument vibrate faster or slower, which changes the sound that's produced.

Trumpet players change notes by pressing the instrument's valves. These extend or shorten the passageway through which air travels. Changing this distance causes the sound's frequency, or rate of vibration, to change. Pressing the valves of a trumpet in a certain sequence produces specific notes—each with its own frequency.

The amount of air Dizzy blows through the mouthpiece also affects the sound's loudness. The more air, the more vibrations, the louder the sound.

If you had been in the room when Dizzy was playing, you may have heard sounds slightly different than what he produced. Sound waves bounce off, go through, or get absorbed by anything in their way. In a room the waves are affected by the walls, floors, and ceilings, just as trees and buildings can alter waves outdoors. That's why listening to music through headphones is a different experience from listening to a stereo or live concert.

Trumpets don't usually bend at a 45-degree angle the way this one does. In the early 1950s, a dancer fell on Dizzy Gillespie's trumpet, knocking it out of shape. Dizzy liked the sound it produced, and from that day on he had all of his trumpets made in this shape.

Wake-Up Call

A rooster SOUNDS OFF

Q: You're a rooster. Must you make so much noise this early in the morning?

A: Of course. I *am* a rooster, nature's own feathered alarm clock, and that's what we do. Roosters crow all day long, by the way, not just at dawn. Besides, I prefer to call it sound, not noise.

Q: Didn't know there was a difference.

A: Sure is. People usually think of noise as something that's annoying . . . distracting . . . confusing, like when your radio is between two stations. Noise isn't distinct and organized, like my own clear, confident call.

Q: So what's sound?

A: Sound is energy transferred by waves. What you people call "sound" is the result of the waves hitting an eardrum, where the vibrations travel to the brain. Your brain tells you it's a sound and what created it. You know my sound: Cock-a-doodle-doo!

Q: Shhh! You can probably be heard for miles!

A: Do you really think so? Great! I just love daybreak because my sound can travel so far, despite the morning chill in the air.

Q: What does the temperature of the air have to do with sound?

A: Sound waves travel better through warm air than cold. And they travel better through water than air at any temperature, but since I can't swim, that doesn't do me much good. But in the morning, there's not much going on out here. That means there aren't many other sound waves to get in the way of my own. So even though it's colder, I can really make an impression.

Q: Where do sound waves come from?

A: All sounds start with vibrations. The sounds I make begin with vibrations in my throat. Strong bands of muscles called vocal cords start shaking and moving when I feel the urge to crow. Vibrations make the air move, and those motions are sound waves.

Q: How does sound form waves?

A: The back-and-forth motion of the vibrations creates waves. It happens as tiny particles of air or water or whatever the sound is traveling through are crushed together, pushed forward, and then fall back. Back, forth. In, out. You can't see the waves, so just think of them as ripples in a pond, spreading out in circles when you toss in a pebble.

Q: So the waves travel out?

A: You got it. Imagine that the pebble symbolizes the source of the sound. The speed at which the ripples are pushed out from the center is called the frequency. The faster those ripples are pushed, the higher the frequency. Like this: cock-a-doodle-doo!

Q: Okay, okay. Then where do the waves go?

A: They spread out in all directions, bouncing off, moving through, or being absorbed by anything that gets in the way—trees, rocks, whatever. All the bouncing can change the sound. Make it lose its edge. Get soft and

dull. Notice that I'm standing up here on top of the roof?

Q: How could I not?

A: I picked this perch carefully. If I went inside the barn and crowed, my sound waves would bounce off the walls and chickens and get absorbed by hay and eggshells. My sound would be muffled. I'd lose most of my audience, which would be a tragedy.

Q: Well, maybe not for all your listeners. All right, I see how sounds are made, but how do people actually hear them?

A: With those two flaps on both sides of your head. Or, actually, because of what goes on inside those flaps. The sound enters your ear and reaches your eardrum. They hit the tight surface of your eardrum and produce more waves. These new waves reach three little bones in your middle ear. Those bones may be small, but they have a lot of power. They beef up the vibrations so they're two or three times as strong. The amplified vibrations now go to your inner ear, which is filled with fluid. That's where they get transferred.

Q: Transferred?

A: Changed. Connected. Charged. We're talking electricity. The inner ear is where the energy of sound is changed into the energy of electricity. Electrical signals that travel along a big

bundle of fibers called the auditory nerve. At last they're ready for the big time.

Q: To be heard?

A: More than that. To be understood. Changed from being just a bunch of electrical signals to a recognizable, familiar sound, one people can pin a name on. Like bow-wow—it's a dog! Or chirp, bleep—it's a fax machine. Or cock-a-doodle-doo—it's you know who!

Q: (sigh) Don't you ever stop?

A: I'm almost finished. All this recognition goes on in your brain. It picks through the sounds and selects the right picture to go with each one. That's not easy, since your brain recognizes about 400,000 different sounds. Then it commands the body to take some action: Let in the dog. Send the fax. Wake up.

Q: Wait. Are you saying that the brain sorts out sounds even when we're asleep?

A: Yep. The human brain is simply amazing. That's why a person can sleep through noisy traffic outside the window and still wake up when the alarm goes off. An electrical, mechanical, or nice feathery alarm that goes . . .

Q: No! Please! Don't make another . . . sound!

A: COCK-A-DOODLE-DOO!

Activity

WHERE, OH, WHERE Explore sound waves with your friends. Find an open space such as a field, and have a few friends gather around you at a distance of about 10 feet (3 m). Put on a blindfold. Your friends should take turns making a sound: clap, snap, shout, or stomp. After each noise, you should point in the direction in which you think the sound came from. Are you correct every time? What would make this test harder?

The Wave of the Present

PITCH

Perfect Pitch

Sound is a form of energy that travels in waves. If you could see sound, it would look like a Slinky™ stretched between two people and pushed backward and forward. Sometimes sound waves move slowly, with a long way between crests.

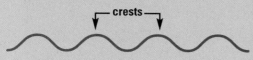

Sometimes the crests come at you thick and fast.

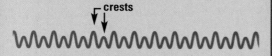

The difference has to do with the pitch of a sound. That's the same as a note, or sound, that's higher or lower than another. If you were to hum the lowest note you can and then look at it, you'd see that the crests on that sound wave are far apart. When that happens, we say the wave has a low frequency.

If you were to hum the highest note possible and could see the sound wave, you'd notice that the crests are a lot closer together. This wave has a high frequency and you are using a lot more energy to create it.

The closer the crests, the higher the frequency—and the higher the sound we hear. Frequency is measured in units called Hertz, which is abbreviated as Hz. To find Hertz, we look at how many crests pass a given point in one second. If 500 wave crests pass a point in one second, then we say the frequency is 500 Hz.

In the Right Range

Four musical voice parts range in frequency.

soprano (high female voice)   tenor (high male voice)
alto (low female voice)   bass (low male voice)

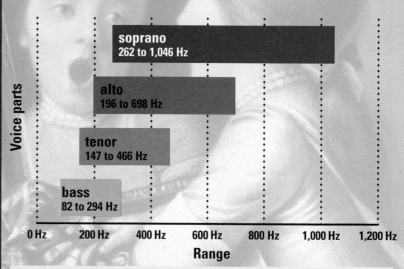

soprano
262 to 1,046 Hz

alto
196 to 698 Hz

tenor
147 to 466 Hz

bass
82 to 294 Hz

Voice parts

0 Hz   200 Hz   400 Hz   600 Hz   800 Hz   1,000 Hz   1,200 Hz

Range

Fun Fact: Dolphins, bats, and grasshoppers can produce sounds more than 100 times higher than those of a soprano singing her highest note.

Hear, Hear!

Humans can hear sounds with frequencies as high as 20,000 Hz—high above anything anyone can sing. Impressive? Sure, but many animals hear much higher frequencies.

Animal	Lowest audible pitch	Highest audible pitch
Human	20 Hz	20,000 Hz
Dog	15 Hz	50,000 Hz
Cat	60 Hz	65,000 Hz
Dolphin	150 Hz	150,000 Hz
Bat	1,000 Hz	120,000 Hz

Helpful Hint: Don't bother singing to bats unless you're a soprano.

The Need for Speed

Suppose you are standing at one end of a football field and your friend is at the other. If you shout your friend's name, the sound of your voice—the sound wave—travels from your mouth to your friend's ears. Do you think a third friend could run fast enough to follow the sound from your mouth to your friend's ears on the other side of the field? No way! On a warm day, sound waves travel close to 740 miles (1,191 km) an hour. Sounds like a lot, right? It is! It's 1,115 feet (340 m), or the length of four football fields, per second.

Break Through

The speed of sound waves also depends on where the sound is traveling. This chart shows how quickly sound travels through substances.

		speed in meters/second
gases	through air	340 m/sec
	through alcohol	1,130 m/sec
liquids	through water	1,230 m/sec
	through gold	3,240 m/sec
solids	through steel	5,000 m/sec
	through granite	6,000 m/sec

Obviously, air is pretty low on the list. But it beats outer space. Outer space is completely empty—we call it a vacuum—and sound can't form at all because there's nothing to carry the waves.

Turn That Thing Down!

The loudness of a sound is measured in decibels, abbreviated as dB. The higher the number, the louder the sound. It is a logarithmic scale in which each 10-point move up the scale is 10 times louder. For example, 30 dB is 10 times louder than 20 dB. A level of 0 dB is a sound that can barely be heard. You'll probably feel some discomfort or even pain listening to anything that's 120 dB or more.

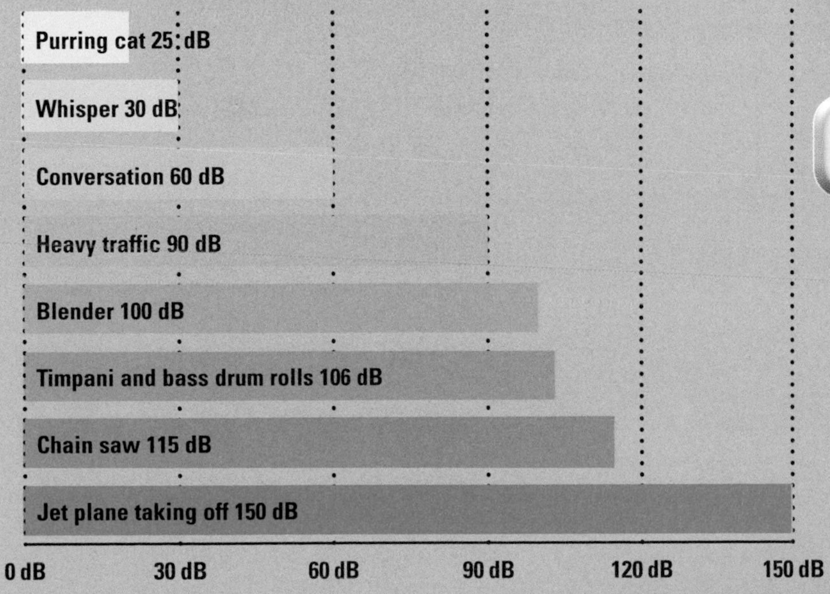

Purring cat 25 dB
Whisper 30 dB
Conversation 60 dB
Heavy traffic 90 dB
Blender 100 dB
Timpani and bass drum rolls 106 dB
Chain saw 115 dB
Jet plane taking off 150 dB

0 dB   30 dB   60 dB   90 dB   120 dB   150 dB

Whatta Pain

Extended periods of listening to noise levels above 150 dB can actually cause deafness. But the decibel level drops as you move away from the source of a sound. A jet engine that blasts your eardrums at 140 dB when you're 98 feet (30 m) away is like 120 dB to someone 197 feet (60 m) away—and a tolerable 105 dB to someone who is 1,968 feet (600 m) from the runway.

Activity

ONE-HAND BAND Stretch a large rubber band between your thumb and index finger. Pluck it with your other hand. What's happening? Do the same activity with a smaller rubber band, perhaps one that's half the size. How does its sound compare to the larger band? Which do you think has a higher frequency? What might the sound waves coming from each look like? What do the results of this activity tell you about an adult's voice compared to a child's?

PLAYING ALONG

What if you could turn yourself into energy, more specifically the energy of a sound wave? To travel with a sound, you'll have to change energy forms a few times. But to experience something like this, it's worth the hassle.

Someone just struck a key on a piano. Sound waves travel out in all directions (right), but a person's ear picks up only a tiny section of the waves. You have only a couple of seconds to turn your flesh and bone into waves of pulsating energy. You're about to fly through the air as a sound wave to the place where all sound waves go—from here to "hear."

TA-DA! Okay, now that you're of "sound" mind and body, let's get through the air. The first pass-over destination is probably familiar. It's the auricle, or outer ear—the fleshy earring holder. This is the collection point for all approaching sound waves.

Hang on! You'll feel like the space you're flying through is getting narrower as the auricle is funneling you past the earwax down the auditory canal.

BOOINNGGG! You just crashed into the eardrum. That happens to every sound wave. The eardrum is the border between the outer ear and the middle ear. You have now changed energy forms—from a wave to a mechanical vibration then for an instant back to a wave. You're about to experience some twists and turns as you hit the three compact bones called ossicles—also known as the hammer, anvil, and stirrup. They make up the middle ear.

The ride gets a little bouncy now. The turbulence is caused by the vibrations you made when you—as the sound wave—hit the eardrum. You have to change back to mechanical energy to get across the hammer, anvil, and stirrup. Those bones act like a drumstick on a drum. Think of the eardrum as the arm of a drummer; you are causing its movement. The middle ear takes the sound waves from the "large" eardrum and compresses them into sounds that can pass through the oval "window" into the inner ear.

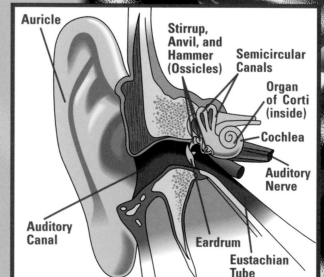

Auricle

Stirrup, Anvil, and Hammer (Ossicles)

Semicircular Canals

Organ of Corti (inside)

Cochlea

Auditory Nerve

Auditory Canal

Eardrum

Eustachian Tube

As you are vibrating through the ossicles, be careful not to make a wrong turn or you could collide with the tonsils. They're at the end of the Eustachian tube, which leads down from the middle ear to the back of your throat. You know how your ears "pop" when you go up in an elevator? That happens because the air pressure in your mouth and nasal cavity changes and presses into your middle ear. Swallowing equalizes the pressure and creates the pop.

Once past the ossicles, you squeeze through the oval window and enter a completely different space. Suddenly you're traveling through a jelly-like liquid. You're spinning inside the cochlea, a snail-shaped bony area. If you don't feel dizzy, thank your own inner ear canals that branch off from the cochlea. The movement of fluid is what gives us our sense of balance.

But balance isn't your problem now. You're trying to get to the brain so you can tell it what sound you are delivering. You're almost there.

The cochlea is getting narrower. As you reach the end of the snail shell, you pass over the Organ of Corti. You'll know it by all those tiny hairs. The Organ of Corti has more than 20,000 hairs, which respond to sounds of different frequency. Here you part company with any other sound waves. As you pass over the hairs, those that are tuned to your frequency will respond. The movement of the hairs turns you into an electrical signal, and you move through one of 30,000 nerve fibers to connect with the part of the brain that "hears" the sound. The brain recognizes your familiar sound. This leads to other neuro-chemical reactions affecting the muscles in the face.

You've reached the end of the journey. The person smiles while recognizing a familiar tune. The sound you are carrying has been heard loud and clear.

Activity

PHONY PHONE CALL Build a phone for you and a friend using two empty cans and string. It may be helpful to get a bar of wax and coat the string with wax. How does the can focus the sound wave? How is the sound wave, or vibration, conveyed to the can? How is the vibration in the can conveyed to the string? Why does the wax help? (Hint: Think of a real phone.) How is this an example of energy transfer? If one of you goes around a corner will the phone still work? Why?

MUSIC, SWEET MUSIC

What sounds like music to one person's ears may be noise to another. But the power of music's vibrations can't be denied.

Tokyo, Japan, January 10, 2000

I am playing now with earplugs because the sheer volume of our rock power was causing me great pain. It is a disorienting effect to get used to the softness of the music with the plugs in, [but] I am starting to become accustomed to it. Hopefully, it will stop me from having to wear a hearing aid within the next few years.

—Flea, bassist for the Red Hot Chili Peppers

It may surprise you that many musicians wear earplugs on stage. But the fact is that years of listening to loud music can take a toll on hearing. Exposure to loud sounds causes damage to the delicate hair cells of the inner ear, which conduct electro-chemical impulses to the brain.

Tinnitus (TIN-uh-tuss), a some-times-painful ringing or buzzing in the ears, is often an early signal of hearing loss. You might hear the ringing in your ears after listening to a concert or your Walkman at a high volume. For most people the problem is only temporary, but imagine the damage you could do if you play an instrument each night and practice two or more hours a day.

To lessen the damage, many musicians wear special earplugs that reduce sounds by up to 25 decibels. The plugs help drown out unnecessary noise so musicians can hear themselves better and save their hearing.

Heiligenstadt, Austria, October 6, 1802

It was not possible for me to say speak louder, shout, because I am deaf. Alas, how would it be possible for me to admit to a weakness of the one sense that should be perfect to a higher degree in me than others, the one sense which I once possessed in the highest perfection.

—Ludwig van Beethoven, classical music composer

Ludwig van Beethoven, one of the world's greatest classical music composers, started losing his hearing at the age of 28. It became progressively worse until he was about 45. Then he could no longer keep his deafness a secret.

Although his deafness caused him to stop making public performances as a pianist and conductor, Beethoven did not allow it to end his musical career. He continued composing throughout the last decade of his life even though he was totally deaf. Amazingly, he used the piano's vibrations, which he felt with a wooden stick. He placed one end between his teeth and rested the other end on the piano.

SIT HEAR!

Hearing loss may have something to do with a musician's seat near the instruments and amplifiers. Rock music drummers tend to have more hearing loss in their left ear, since that's the ear closest to the shrill high-hat cymbals (circled, right). Classical musicians, usually surrounded by other instrumentalists, are also affected by where they sit. Violinists usually have more problems with the left ear's hearing, while flute and piccolo players experience greater loss in the right ear.

THE KEYS TO PIANO PLAYING

Tapping a piano's keys while pushing down the sustaining pedal—the one on the right—with your foot allows the strings inside the piano to continue to vibrate. This makes the notes last longer, even after you let go of the keys.

Fine Form

The solid body of an electric guitar doesn't affect the sound that is heard, so it can be any shape or size. In 1958 guitar manufacturer Gibson created the futuristic Flying V (right) to boost guitar sales.

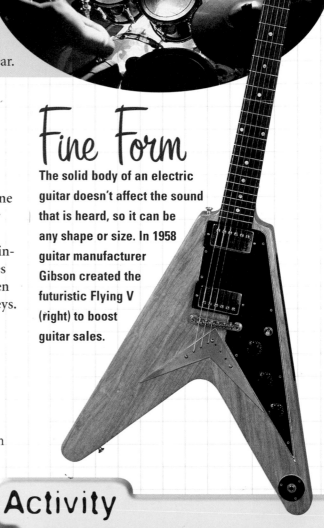

Stringing It Along

Instruments, such as the violin (left) and the guitar, have stretched strings that vibrate to produce sounds when tapped, plucked, or bowed. The pitch of the sound produced depends on the length of the string, its thickness, and how tightly it's stretched. The vibrating string is lengthened or shortened depending on where musicians press with their fingers. That's how they can get so many notes from so few strings!

VOICES CARRY

When you sing, your vocal cords—tight bands in the throat—vibrate as air passes over them. Throat muscles can change the tightness of your vocal cords to produce different notes. Women usually have higher voices than men because their vocal cords are shorter.

Activity

MUSICAL GLASSES Line up eight drinking glasses—four large and four small—and pour different amounts of water into each. With a pencil or metal eating utensil, strike each glass near the top. Note the different tones produced by the each glass. Experiment with the amounts of water and the order of the glasses until they begin to sound like the notes on a musical scale—lowest to highest pitch. How can you scientifically explain what you are hearing? Why does the sound change depending on how much water is in a glass? Create a chart that lists the amounts of water versus the order of the tones.

Catching the VIBE

1600s	1800s	1872	1900
The ear trumpet (below), a large horn-shaped device intended to act as a "sound funnel," is introduced in Europe. This simple aid gathers sound and directs it into the ear.	King Goa VI of Portugal orders ear trumpets inserted into the armrests of his throne. Others find ways to conceal the trumpets within the fashions and furnishings of the day, including in canes, fans, and umbrellas.	American inventor Alexander Graham Bell creates the first non-horn-shaped hearing aid. Bell made this labor of love as a gift to his hearing-impaired mother.	The first electric hearing aids (below) are produced. These devices are large and provide little help to those with severe deafness. Acoustic fans are also used. The thin rubber fans conduct sound vibrations through a person's bones. Users put the fan between their teeth and bend it toward a sound. The sound strikes the fan, sending vibrations through the user's teeth to the skull and finally to the auditory canal.

As society and industry advance, our surroundings are growing noisier. Hearing aids are becoming a necessity for more people than ever. Fortunately, over the centuries we've learned how to improve technology to develop mechanical devices to help the hearing impaired.

1930s

Hearing aids become more compact. In the early 1930s, they are made with smaller tubes than previous models, microphones, and two batteries.

1950s

The invention of the transistor—a device that amplifies sound—paves the way for smaller aids that can be placed behind the ear and in eyeglasses. The tiny transistor replaces tubes in earlier hearing aids. Newer hearing aids include a microphone, amplifier, battery, and an ear mold that connects the system to the ear.

1970

The first cochlear implant is inserted into a person. Tiny wires implanted into the middle ear serve as an artificial cochlea. An external microphone is connected to the cochlea through a plug behind the ear. This cochlea transmits electrical impulses to the part of the brain that translates impulses into sounds.

1990

The Natura (above), the world's smallest digital hearing aid, is launched. It features the world's smallest microchip, which can be programmed with a handheld computer. Like mini-computers in your ear, digital aids can "think" about sound. They tone down distracting background noise, such as air conditioners, and enhance the voices of people nearby.

Helen Wilkens, 10, wears a hearing aid in 1952.

Activity

HEAR OR THERE? Experiment to see what sounds your pet can hear and how it reacts to different pitches. Keeping in mind that its hearing is usually more sensitive than ours, perform three varied sound effects for your pet, standing at least 5 feet (1.5 m) away. Keep your performance short—a few seconds each—standing on opposite sides of the pet each time. Then move a foot backward and repeat the test until you are 15 feet (4.5 m) away. Record your observations in a table format, and make an attempt to explain each outcome. Share your observations with the class and combine findings for each kind of animal, in an attempt to make general theories about animal hearing.

WIRED FOR SOUND

Imagine yourself in a room with just a few close friends, talking and laughing. The conversation is . . . too swift for you to follow on their lips, too difficult for you to understand because you are deaf. —Beverly Biderman

If you couldn't hear, this would be a familiar experience for you. You might find it difficult and isolating, but you could find ways of dealing with it. Beverly Biderman (left) had to. She was born with a hearing impairment. For most of her life, she did a remarkable job of passing as a hearing person. Beverly's parents didn't even discover her disability until she was four. Her hearing became progressively worse as she got older. By the time Beverly was 12, she could no longer use the telephone or listen to the radio.

At the age of 46, Beverly decided to take advantage of a breakthrough in medical technology to see if it would improve her hearing. She was one of only 20,000 people worldwide to receive a cochlear implant, the first effective artificial sensory organ for humans. This is her story.

After the second or third grade at school, I started to get into trouble. . . . I was disruptive [because I did not hear what I was disrupting], and noisy [because I could not monitor the loudness of my voice]. . . . Our family doctor suggested a trip to a specialist.

Beverly was fitted with a hearing aid. It helped, but her hearing continued to worsen.

Unlike a cochlear implant, which allows people to hear more sounds, Beverly's hearing aid amplified everything on the chance that her remaining hearing might pick up some of the loud sounds. Yet Beverly's hearing was so poor that without the hearing aid, she would have heard no speech sounds at all.

Beverly used several strategies to cope with her hearing impairment.

The most important of these was lip-reading or, as it is sometimes called, "speechreading," because it involves more than the lips. If people wore sunglasses, for example, I would have trouble understanding what they said, as I needed to see the expressions in their eyes. . . . I would startle them by saying, "Please take off your sunglasses. I can't hear you."

Only one-third of the sounds that words are made of are "readable" on the face. Beverly filled in the blanks by considering the context of what was being said and following the rules for how sounds and language works. She also relied on "feeling" sound through vibrations.

Time for a Change

In 1992, Beverly's doctor suggested she get a cochlear implant in one ear. Cochlear implants directly stimulate the auditory nerve fibers, which send signals to the brain.

David [my audiologist] brought out an implant system, and showed me the parts of it. It had a gray processor the size of a cigarette pack that I could wear on a belt, or in a pocket. . . . The microphone was set in a beige earpiece that I would wear over my ear much like a behind-the-ear hearing aid. This microphone would pick up sounds in my environment and send them down to the processor through a thin beige cable I would wear under my clothing.

Six weeks after the implant was installed, the system was turned on. This was the most anxious day of Beverly's life. She was "wired for sound," but had no idea of what or how well she would hear.

Cochlear implant: How it works

1 Microphone picks up sound.

2 Sound goes from microphone to separate speech processor.

3 Processor translates sound into electrical codes.

4 Codes travel a transmitter.

5 & 6 Transmitter sends codes across skin to receiver/stimulator.
Receiver/ stimulator converts codes to electrical signals.

7 Signals go to electrodes in cochlea to stimulate neurons.

8 Neurons send messages along auditory nerve to the central auditory system in the brain, where they are interpreted as sound.

[David] gives me a little pep talk about how at the beginning, some people find that voices are mechanical-sounding, like cartoon-character voices. . . . He then turns on all the electrodes and asks, "How does that sound?". . . David's voice, when I realize it is a voice . . . sounds like . . . horns and whistles.

David patiently walks me around the clinic to have me listen to all the torturous environment sounds: He runs water in the sink, and it sounds like Niagara Falls; he crumples up a plastic bag, and it makes a harsh crackling sound.

A New World

While Beverly took time to adjust to her new hearing, she eventually discovered the differences the implant made in her life.

When I get outside . . . there are what seem to be thousands of birds, all very noisy, but the sounds together are rather nice. I feel connected, a part of the scene. . . . My feet scuff against the gravel path, and the sound is satisfying and pleasant.

By one year . . . I started to use the phone regularly and began to have some good long conversations with friends and family who were willing to speak slowly.

The implant provided Beverly with experiences she could never dream of. Now a writer and speaker, she relates ideas and stories about deafness, cochlear implants, and adapting computer technology for the deaf.

Activity

HEARING IS BELIEVING You don't have to be born with a hearing impairment to lose your hearing. At any age, exposure to loud sounds over time could permanently affect your hearing. A person your age could find themselves in Beverly's shoes.

What would that be like? Find three separate half-hour blocks in the course of a day to block out sound—perhaps getting ready for school, at lunchtime, and after school. During these periods, insert ear plugs (available at most drug stores) or ear "head sets" like those wore by airport personnel on a runway. These devices will block out about 20-40 dBs of noise, so you will hear some muffled sounds. Write a diary entry about what it was like to go without hearing. What did you miss? How did people react to the fact that you couldn't hear? Finally, think about how your daily life would be different without the ability to hear, and how you can protect it. How do you think your Walkman, TV, and concerts affect your hearing over time? Create five suggestions on how to take care of your hearing.

CLEAR AS A BELL

Faxes, cell phones, the Internet—where would we be without Alexander Graham Bell's invention of the telephone?

When Bell was a young man, the telegraph was still a marvel. Invented in 1837 by Samuel F. B. Morse, the system transmitted 15 to 20 "words" a minute across a wire. Yet no voices could be heard through this device. Words could only be transmitted through Morse code—a system of short and long electrical pulses, known as dots and dashes. Plus, the telegraph was costly and slow. Bell was determined to improve it.

Unlikely Inspiration

In his mid-20s and eager to create an invention that would outdo the telegraph, Bell had a friend translate a German book about producing and transmitting sound. In it, the author described a device that produced vowel sounds using tuning forks, electromagnets, a battery, and a resonant chamber. From the translation he was given, Bell mistakenly understood that these vowel sounds could also be transmitted by wire. And if vowel sounds could be transmitted, he reasoned, why not other sounds as well?

While the idea of transmitting speech electrically was planted in Bell's mind, his attention was focused on the telegraph. He wanted to create a "harmonic telegraph" that could send several messages at one time. Bell figured that by sending messages at different pitches, they all could be sent simultaneously without getting jumbled.

The Magic of Teamwork

Bell teamed up with Thomas Watson, who had electrical, mechanical, and model-building skills. On June 2, 1875, Bell and Watson were working in separate rooms in an electrical shop in Boston, Massachusetts, when Bell heard a twang that Watson had accidentally generated. The sound was complex, similar to the human voice. It convinced Bell that conveying speech over a wire was possible. How lucky it was, said Watson, that "the right man had [a sound sensitive] mechanism at his ear during that fleeting moment."

The telegraph had utilized an intermittent—on and off—current. It was good for conveying short and long electrical pulses of Morse code, but not the complex rising and falling tones of the human voice. "If I could make a current of electricity vary in intensity precisely as the air varies in density during the production of sound," Bell wrote, "I should be able to transmit speech telegraphically." Bell and Watson constructed a new device that had a conducting liquid that could change its resistance and produce a continuous current that could rise and fall.

The proprietors of the Telephone, the invention of Alexander Graham Bell . . . are now prepared to furnish Telephones for the transmission of articulate speech through instruments not more than twenty miles apart. Conversation can be easily carried on after slight practice and with the occasional repetition of a word or sentence. . . .
—a telephone advertisement from May 1877

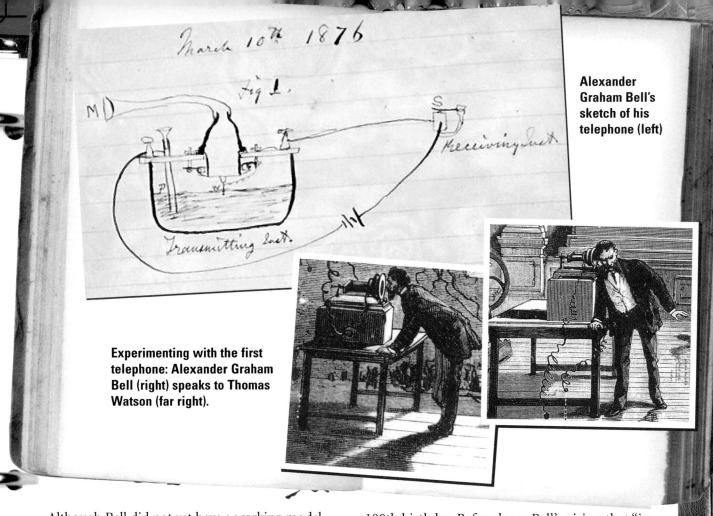

March 10th 1876

Fig 1.

Transmitting Inst.

Receiving Inst.

Alexander Graham Bell's sketch of his telephone (left)

Experimenting with the first telephone: Alexander Graham Bell (right) speaks to Thomas Watson (far right).

Although Bell did not yet have a working model of a telephone, he now understood enough to transmit voices telegraphically. He applied for a patent so that no one could copy his ideas. Remarkably, Bell filed his application only hours before his chief competitor, inventor Thomas Edison!

Sounds Like Team Spirit

Bell and Watson persisted in their experimentation, and on March 10, 1876, the groundbreaking moment arrived. "I then shouted into M [the mouthpiece] the following sentence: 'Mr. Watson—come here—I want you.' To my delight he came and declared that he had heard and understood what I said." Alexander Graham Bell had just made the world's first telephone call, and Thomas Watson had received it.

A few months later the telephone was introduced at the Centennial Exhibition in Philadelphia, the world's fair celebrating America's 100th birthday. Before long, Bell's vision that "in the future a man in one part of the country may communicate by word of mouth with another in a distant place" became a reality.

When Bell passed away on August 2, 1922, telephone service throughout the country was stopped for a minute of silence.

Activity

BLOCK AROUND THE CLOCK Hold a clock to your ear and listen for the ticking sound. Slowly move the clock away until you no longer hear it. Measure the distance at which you can no longer hear the clock ticking, and compare your measurement with other students'. Is it the same? Why or why not? Now place a long cardboard tube—at least as long as a paper towel tube—against your ear, and hold the clock up to the other end. How does what you hear compare with the sound the clock makes without the tube in between? Knowing that sound needs energy to travel, how might the tube have affected the waves? What does that mean in terms of you hearing a sound? How could this explain the shape of horn instruments and the telephone?

Now Hear This

Sounds are everywhere. So scientists versed in sound-related topics are hard at work studying them. Here are some places where interesting observations in sound research have recently been made.

West Lafayette, Indiana

Engineers at Purdue University have found that highway noise is harmful to nearby residents' health. The engineers are improving highway barriers that can cut down on noise and create guidelines for car noise emissions.

North America

Nashville, Tennessee

Laughter causes different reactions in listeners, depending upon its tone and pitch. Researchers at Vanderbilt University claim that people usually respond positively to high-pitched laughter with songlike qualities.

South America

Los Alamos, New Mexico

Sound waves are used to pinpoint damage in structures after earthquakes. Researched at Los Alamos National Laboratory, the technique measures sound waves as they pass through structures, monitoring changes in sound tone and resonance. The waves show weak spots and cracks in buildings, bridges, and pipelines.

University, Mississippi

Researchers at the University of Mississippi have developed the Aqua-Scanner™, a device that uses sound waves to track the number and size of catfish in shallow ponds. The longer the sound waves are, the bigger the fish.

Chichen Itza, Mexico

Clapping in front of the staircases of an ancient Maya pyramid produces an interesting echo. Scientist Daniel Lubman claims the Maya constructed the pyramid so that its echo would sound like the quetzal, a sacred bird.

Greenland Sea

Knowing there is a relationship between the speed of sound and changes in temperature, scientists are using sound waves to study the average temperatures of this body of water. By measuring the time it takes sound to travel between two locations, they can calculate the sound's speed.

Nagu, Finland

Researchers working on the World Soundscape Project have found that the soundscape in Nagu changes with the seasons and affects residents' moods. Summer offers a loud soundscape, with bustling tourism and rolling waters, creating tense behavior. Winter brings quieter surroundings and calmer moods.

 Europe

Asia

Africa

Papua New Guinea

The echoes used in the songs sung by the Kaluli people of New Guinea represent sounds commonly heard in their rain forest environment, according to scientist Steven Feld.

Australia

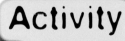

Black Sea

Researchers from the Tubitak Marmara Research Center are mapping the sea floor by bouncing sound waves off it and noting how the waves react. These scientists are finding fault lines and other imperfections, furthering our understanding of earthquake activity and the composition of Earth's crust.

Amazon Basin, Brazil

Researchers from Wild Sanctuary, Inc., recorded the area's morning sounds to study bird and insect behavior. A low-flying jet disrupted the recording and altered the kinds of sounds the birds and insects made. This incident shows that mechanical noise and the loss of natural sound has a major impact on living things, from increasing stress to changing behavior.

Activity

SOUNDS LIKE RESEARCH Create a pin-drop test for your friends or family members of different ages. Ask each of at least five people to stand backward approximately one foot (30 cm) away from a table. Drop the pin on the table from a height of 5 inches (13 cm) and ask the person to raise a hand if he can hear it. Measure the distance between the person and the table. Have the person move one foot away from the table and repeat the test, dropping the pin from the same height each time, until the person can no longer hear the pin drop. After testing, create a data chart of the results, showing distance dropped versus age of participants. Are the results different from person to person? What may explain this?

O'Ears to Them!

Have you heard? All ears don't look and act like ours. Still, lots of animals keep their ears busy doing important things.

6.5 feet

Big Shot African elephants have the biggest ears on Earth. From top to bottom, their giant flaps can measure 6.5 feet (2 m)—almost as tall as a pro basketball player. The size of an elephant's ears makes them helpful receivers of sound. They can pick up calls of other elephants more than 5 miles (8 km) away.

See-Through Sea Life If you look closely at this transparent swimmer (right), you'll see a bloated kidney bean-shaped part, called an air bladder. The glassfish collects sound waves using its middle ear—where the air bladder is located. A lateral bone parallel to its backbone connects the air bladder with the fish's inner ear. The air bladder vibrates and transmits sound waves to the fish's inner ear.

Feel The Beat Have you ever seen a movie with a snake charmer playing music? A snake rises out of a basket and "dances." The snake is actually following the movement of the horn, rather than swaying to the music. Snakes can hear sounds, but not the way we do. They don't have an outer or middle ear, so they need to feel the music, or the rhythm of approaching preys' footsteps. A snake's jawbones detect vibrations made by movements. The vibrations travel through other bones to the inner ear, where they are translated into signals that the brain interprets as sound.

Simple Sound If you were to pretend to listen like a cricket (right) in class, you'd have to point your knees up at your teacher. The ears of this insect are on its abdomen, behind its front legs—and they have to be aimed at the sound the cricket is trying to hear. Like many other bugs, these ears are simple: just an eardrum and nerves that pick up vibrations. It seems basic, but a cricket's ears can often detect sounds that are out of human range.

Underwater Amplification To find the ears of a killer whale, just look for the quarter-inch slits on both sides of its head. Inside the skull, the whale's inner ears are similar to ours—just a lot bigger. One major difference: The whale's middle and inner ears are set in a thick bone called the tympanic bulla. This location helps intensify underwater sound waves. Since water is more dense than air, the tympanic bulla gives the "dulled" underwater sound a little boost as it travels through the auditory canal to the whale's brain. Like bats, whales use echolocation to navigate their often murky environment.

Activity

GUESSING GAME With a friend, practice bouncing a tennis or hand ball off the wall of a handball court or building. Get a sense of how long it takes to hear the ball hit the wall when you're standing 10 feet (3 m) away, then 20 feet (6 m), then 30 feet (9 m). Put on a blindfold and have your friend walk you to different "secret" locations away from but facing the wall. To mix things up, the person may walk you in a circle or back and forth before putting you in a spot. Now throw the ball against the wall. Based on the amount of time it takes before you hear the sound, estimate how far you are from the wall. How is this experiment similar to echolocation? How is it different? What would a person need for echolocation to work?

THE CASE OF THE Collapsed Cards

"Someone is sabotaging my house of cards!" shouted Buzz Brattle as he breezed into his room after soccer practice.

The evidence: 36 playing cards scattered across his desk.

For the third day in a row, Buzz constructed an elaborate mansion by carefully stacking cards. He and his sister, Melody, were each competing for the Brattle record, which their brother Crash currently held at 37 cards.

For the past two days, Buzz had come home to the very same scene of scattered cards. The first day he thought it was the wind, so he closed the windows in his room. When it happened again the second day, Buzz thought someone in his family might be playing a joke on him. So he planned a way to catch the culprit. He set up a tape recorder and microphone to record the sounds of anyone who entered his room. Today, before soccer practice, Buzz popped in a 120-minute tape, just the right length to record all the sounds while he was away. Finally, he'd be able to prove who was guilty.

Buzz pressed "rewind" on the tape recorder. As it rewound he decided to question the suspects.

He stormed into Crash's room to find him putting sheet music away. "Did you knock my cards over?" Buzz asked accusingly.

"No way. I've just been practicing my guitar," Crash said.

"Hey, is something burning?" Buzz asked.

"No. I made a burger. Burnt it a little, I guess, 'cause the stupid smoke alarm went off," replied Crash. "Hey, does this mean I win? Take this dish back downstairs. Man, I'm the winner and that means you should be doing my chores for a week."

"Not if I can help it!" Buzz grumbled, closing the door.

In the kitchen, his mom was peeling vegetables.

"Mom, did you notice if anyone went into my room while I was at soccer practice?"

"Well, let's see . . . I was outside gardening while you were out and your brother was practicing. Oh, I did put some clean laundry on your bed just after your Aunt Mabel called."

"Did you see my cards?" Buzz queried.

"Yes! You've stacked them up quite nicely, dear."

"I went in your room," piped in Melody, who was sitting at the kitchen table thumbing through a magazine.

Buzz eyed his sister—and rival—suspiciously.

"I was looking for that Sonic Boomers CD that you borrowed the other day," Melody explained. "I desperately needed to listen to something besides Crash's guitar playing."

"What's all the ruckus about?" asked Buzz's dad, joining the others in the kitchen.

"I'm just trying to figure out who knocked over the house of cards that was standing on my desk against the wall," said Buzz, starting to wonder if he'd ever succeed in beating Crash's record.

"I'm afraid I can't help you, son. I popped in your room to say hi when I got home. Tripped up the stairs on my way up—Honey," he said, glancing at his wife, "we have to fix that top step. Anyway, son, you weren't there yet. But I did notice that your construction had already collapsed."

Buzz took note of what everyone said. Then he compared each person's testimony with the sounds he recorded while he was at soccer practice.

After listening to all the sounds, Buzz realized who was responsible for his wrecked card house.

Based on the illustration at right and the clues below, can you determine who the guilty party most likely is and why?

Buzz Brattle's House

Buzz's room

Melody's room

Crash's room

Use these clues to figure out who knocked down Buzz's house of cards.

Clues

*Hint: sound waves can pass through different media, including walls and water.

ELAPSED TIME (hours/minutes/seconds)	SOUND	SOUND LEVEL
0:15:28–0:15:39	telephone ring (3 times)	loud
0:30:01–0:30:06	footsteps	soft, clear
0:30:07–0:30:08	muffled *thud*	soft
0:30:09–0:30:10	"Achoo!"	loud
0:30:09–0:30:14	footsteps	soft, clear
0:42:23	door slam	soft, muffled
0:42:29	alarm rings	sound level rises to same level as 1:03:36
0:43:04	"Hey Mom!" (male voice)	loud
1:03:36–1:30:11	electric guitar music	muffled, then very loud (almost as high as 1:03:36)
1:10:00–1:11:00	thud, "Ouch!"	soft
1:13:00–1:14:00	door click	sound level drops slightly
1:12:21	door click	loud
1:12:22–1:12:26	footsteps	no change in sound level
1:12:23–1:15:01	*sliding, shuffling, clicking,* "Here it is!"	no change in sound level
1:15:02–1:15:06	footsteps	soft
1:15:06	door *click*	faint
1:53:58	car door slam	

Good Vibrations

You know that sound can be a powerful way to communicate, but did you know that it's also strong enough to save the world? Well, that's an exaggeration, but it can do extraordinary things.

REAL ROCK MUSIC

It was the loudest rock concert ever and it was produced by one of the oldest performers. Indonesia was rocked by the Krakatau (crack-ah-TOE-ah) volcano in August 1883. It erupted over a 36-hour period, after keeping silent for more than 200 years. People reported hearing this volcano's blast more than 2,000 miles away in Australia. The final explosion on August 27th is considered the loudest sound created in modern times. Every sound collecting instrument in the world at the time recorded the sound of the blast, and some recorded it all seven times it circled Earth.

SHATTERING SOUNDS

Do you think someone could sing at a pitch high enough to shatter glass? Though it is difficult, it's possible. Well, sort of. While you shouldn't do this at home, here's what it takes: Tap a fine crystal glass with a rubber mallet to make it vibrate and determine its natural frequency of vibration. Then have a person sing in tune with the glass, holding the note. Amplify that sound to around 94 decibels. This should cause the glass to vibrate and shatter.

There are stories of opera star Enrico Caruso and others shattering glasses with their voices, but there is no record of an unamplified voice ever accomplishing this feat. Humans simply don't have the vocal ability to create such a sound.

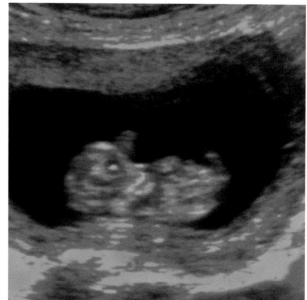

Vision Quest

Who knew that sound could help you see? Doctors and scientists routinely use ultrasound to see inside a person's body for medical procedures. Ultrasound is used for a variety of things, including determining the size, sex, and health of an unborn child, assessing the health of a patient's heart, and identifying tumors and blood clots.

The technique involves moving a microphone-like device over a part of the body. This device emits sound waves that bounce off tissues and organs inside, creating echoes that are fed into a computer. Based on the strength of the echoes, the computer creates an image.

Injections Without Needles

Squeamish about needles? New ultrasound technology may soon take care of your fears. The technology was developed for people with diabetes, a condition that affects metabolism. Some of these people need to prick their finger three times a day to measure glucose or sugar levels in their blood. The ultrasound technique allows people with diabetes to put low-frequency ultrasound waves on their skin for a minute to make a small opening that will stay open for up to 15 hours. They put a patch over this opening to monitor their glucose levels throughout the day. They can use the same technology to inject insulin, the hormone that is lacking in diabetics.

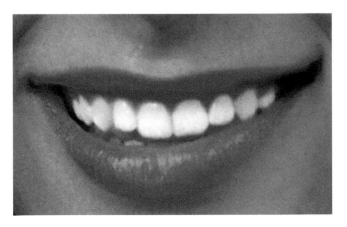

A Sound Cleaning

There's a new sound in teeth brushing: silence. Sonic toothbrushes look similar to electric toothbrushes, but they use bursts of ultrasound waves to knock off plaque. The sound waves also go deep in-between teeth and over gums to kill bacteria. Though you can't feel the sound waves working—and you certainly can't hear them—they do an excellent job of cleaning places in your mouth that you couldn't reach otherwise.

Doctors use similar technology to break up kidney stones, hard buildups in the kidneys formed by calcium salt deposits. These "stones" range in size from microscopic to golf ball.

Activity

FUTURE SOUND Doctors and scientists use ultrasound to find hidden objects and break apart hard to reach obstructions. Do you think ultrasound could be used to find buried treasure? Could it be used to help find life on other planets? Could it be used to predict earthquakes? Think of some ways in which ultrasound can be used. Why are these uses important? In a written explanation, describe how the properties of sound are at work in the uses you've imagined.

Supersonic Soarers

Have you ever seen a severe thunderstorm? You probably saw a flash of lightning but didn't hear the accompanying thunder until seconds later. That's because sound travels almost a million times slower than light. While the speed of sound passes through air at an average of 740 miles (1,191 km) per hour, the speed will change depending on air temperature and materials it passes through. At sea level, sound travels at about 1,100 feet a second. Several miles above our heads, the weather is bitterly cold, and the speed of sound is slower. At these higher altitudes, the air is colder and thinner. Sound waves up there slow to about 660 miles (1,062 km) per hour.

When an airplane flies *faster* than the speed of sound, it creates an explosive sound called a sonic boom. As an airplane approaches this speed—called Mach 1—it catches up with the sound waves traveling in front of it, compressing the atmosphere. This creates incredible pressure in front of the plane that acts as an invisible barrier—called the sound barrier—that the plane must break through.

As the airplane passes through the sound barrier, it pushes molecules of air ahead of the sound waves. These molecules crash into the surrounding air, which in comparison is hardly moving. The result: A thunderous sonic boom and massive shock waves—which can cause serious damage, such as cracked walls and shattered windows—rippling behind the airplane.

Chuck Yeager's plane

Faster than the Speed of Sound

In the 1940s pilots flew planes faster than ever before, almost reaching the speed of sound. But as they got close to Mach 1, they experienced turbulence so great that sometimes it tore the wings off planes. Many people believed pilots would never break the sound barrier. But in 1947, American pilot Chuck Yeager proved otherwise. He flew through the sound barrier at 670 miles (1,078 km) an hour in a specially made aircraft.

Over the next five years, two women tried to become the first to fly faster than Mach 1. In the process they made "sound barrier history."

An F/A-18 Hornet aircraft

A photo taken with a special camera shows an F-18 plane breaking the sound barrier. Shock waves surround the plane.

Jacqueline Cochran

Jacqueline Auriol

New York, 1933

As a child, American Jacqueline Cochran dreamed of soaring to great heights. At age 27, Cochran had her pilot's license and was breaking speed records.

Cochran was the first woman to break the sound barrier. She achieved this by flying her plane in a vertical dive. She described her experience.

I was hanging face downward diving at Mach 1 with my blood surging into my brain, not looking forward to what pilots call a 'red-out' [the possibility of blood vessels bursting]. . . . Shock waves look like rain. [It's like] flying inside an explosion . . . . You hear just fine up there, but you don't hear the sound of your plane. That sound passes behind you because you are going faster than sound can travel.

Spectators on the ground heard the explosive sonic boom. A movie studio tried to film the event, but you'll never see the flight. The boom broke the camera lens.

Cochran flew faster in each consecutive flight, and she succeeded in winning more aviation awards than any other person in history.

Paris, 1949

By age 31, Frenchwoman Jacqueline Auriol had received her private pilot's license. One day while flying as a passenger, her plane crashed. She survived, but suffered a fractured skull, broken collarbone, and nearly lost her arm.

Yet after more than 30 operations, Auriol vowed she'd keep flying and try to break the sound barrier.

In 1953 Auriol crossed the sound barrier for the first time—the second woman in history to do so. Like her friend and rival, Jacqueline Cochran, Auriol dropped the plane into a vertical nosedive to achieve the feat. She described her experience.

The invisible sound barrier really was a barrier: something hard to penetrate. The plane shuddered and shook and vibrated, it went over on one wing. . . . I watched my Mach-meter—when the needle passed the 1, my earphones were filled with shouts of joy. It was my colleagues in the monitoring room who had just heard the sonic boom. . . . These bursts of joy took me by surprise, for there in my narrow cockpit I could not hear the sonic boom. [But] I could see the ground rushing up towards me.

Auriol went on to set many supersonic speed records, earning her the French Légion d'Honneur and prestigious aviation awards from U.S. Presidents Harry S. Truman and Dwight D. Eisenhower.

Awesome Audibles

F-owl Looking

Owls don't have matching ears—one is lower than the other. This allows them to sit high in trees and detect a sound from below in one ear a split second before the sound hits the other ear. It helps them zero in on their prey.

Sound Affects the Movies

Movie scene: A baseball player hits a long drive and starts to round the bases. As he does, the outfielder trying to catch it crashes into the left-field wall, breaking his leg.

What you hear: It's as vivid as if you were right there. The snap of the bat against the ball—WHACK! The crash into the wall—SMASH! The breaking of the bone—SNAP!

But how can the microphones used in filming pick up all that noise? They can't. They use sound effects. A short piece of a rubber garden hose struck by a large piece of bamboo sounds like a bat slamming into a ball. The crash could be made by dumping a collection of metal scraps into a large tub. The broken bone might be a stick or rod surrounded by a layer of paper and cracked in half. And all this is done by sound-effects specialists in a recording studio, after the actors are done filming their scenes.

Classic Queries

If a tree crashes to the ground and there's no one to hear it, does the sound exist?

There may be no brain around to sense the "signal," but there would be a sound because a sound is really just a disturbance in the air. The falling tree will create soundwaves no matter who is—or isn't—around.

Why do you hear the sea when you hold a conch shell over your ear?

The shell's folds vibrate, resonating and amplifying the sounds around it. But what you are hearing are the amplified sounds from the air around you, not the waves breaking against the shore.

Can deaf people dream that they hear voices?
Yes, studies show that even deaf people born to deaf parents can distinguish speaking sounds in their minds. Although their sound representations for words may be different from a hearing person's, they create sounds in their minds as they learn to read.

Moose + Train=Love?

Train whistles once had a much lower pitch. However, they were made higher because the whistle's tone attracted moose to the tracks of approaching trains. Scientists believe the confused animals thought the sound was a mating call from another moose.

SOUNDS LIKE A PROBLEM

In the past statistics showed that people over the age of 65 were the most likely to suffer hearing loss. Approximately 10 percent of people 40 to 65 years old acknowledged some hearing loss. Today the number of people with hearing loss is sharply increasing because sound is a bigger problem than ever. From stereos and headphones, to leaf blowers and heavy traffic, to sporting events, and even vacuum cleaners— sound is becoming more fine-tuned and louder as our technology advances, creating more machines that are louder than ever before.

The Sound of Silence

For the ultimate listening experience you need a dead room. How can a space be dead? Think technically. An anechoic (an-eck-O-ick) chamber is a room totally without *acoustical reverberations*—echoes, or sound energy wave vibrations bouncing between walls. Anechoic means "without echo." Sound waves in an anechoic chamber are fully absorbed, so no one on the outside can hear them, while a microphone inside could pick up pure, unreflected sound.

An anechoic chamber is as close as you can get to complete silence. That's why loudspeaker manufacturers use such a room to test the sound quality of their products.

The secret behind an anechoic chamber is that it's surrounded by material that absorbs sound, usually fiberglass wedges of different depths (see picture above). Even the chamber's microphones and speakers, which pick up and bring sound into the room, are covered with fiberglass.

How can the room be silent if people are walking in? One speaker manufacturer built a floor made of woven stainless steel that sits 6 inches (15 cm) above a fiberglass floor. It is so effective at muffling sounds that you can't hear any footsteps. "Some people are very uneasy walking into the room," said Roger Russell, a former director of acoustic research at the company. "It's like being up in a hot-air balloon without the burner going. Completely quiet." Sometimes, Russell says, the room's silence can feel so intense that people have to leave. Sounds like anechoic chambers don't give everyone "good vibrations."

Pollution Solution

Everyone is concerned with the state of our planet—conserving natural resources has become a big focus for some politicians. But forget the physical earth for a moment and think about the mental health of the land's people. Sound extreme? Not really. Sound, after all, can have major effects on human and animal personality.

A coalition of health-related organizations, led by the National Institute on Deafness and Other Communication Disorders, have instituted Wise Ears!, a campaign to educate the American public about the dangers of noise pollution. According to the most recent research, 10 million Americans have already suffered irreversible hearing damage from noise—whether it is in the workplace, at home, on city streets, at music concerts, or elsewhere.

Your mission: Join with two other classmates to become a local political team running for office. Everyone should help with all jobs, but you can break up the work so each of you is in charge of a key position. One of you should be the "mouthpiece," or politician. Another can be the chief writer, or press representative. And the other teammate should be the main researcher.

Your platform will be "controlling noise pollution in the new millennium." First choose an area of focus: your state, city, town, village, or school district and an "office" for which to run. Then begin to research—both on your own, through various media, and in interviews with residents of the area. What sounds fill the area? How do the residents feel about those noises? How do they deal with the sounds? What are the real trouble spots in your area? You may want to make a map of the location and pinpoint the areas that need the most help. You should also research the effects of such sounds on people and animal behavior. Contacting Wise Ears! (www.nih.gov/nidcd/health/hb.htm) or other hearing-related organizations may be helpful for background information.

Once you know the problems, find solutions. Expand on residents' solutions for dealing with the sound and research your own ways to cope with noisy elements. How can you come up with technologically correct ways to decrease sound's effects?

After your research is done, bring it to the people. Focusing on the noise problems people seem most concerned with, prepare a 5 to 10–minute presentation and create a poster series that outlines and tries to solve at least five noise problems and offers solutions. To educate the community, you should include a discussion of the physical properties of sound in your speech and posters. Remember, you want everyone's "votes," so be sensitive to the noisemakers, too.

Ready for the ultimate challenge? Enter this or any other science project in the Discovery Young Scientist Challenge. Visit *discoveryschool.com/dysc* to find out how.

ANSWER Solve-It-Yourself Mystery, pages 24–25:
Buzz's house of cards most likely collapsed while Crash was playing his electric guitar. The volume on the amplifier was turned up so high that the sound waves caused the wall against Buzz's desk (and other objects) to vibrate. As Crash increased the volume on the amplifier, the amplitude, or wave height, of the sound waves increased. Sound waves with greater amplitudes have more energy. And the energy of these vibrations was strong enough to cause Buzz's cards to fall.

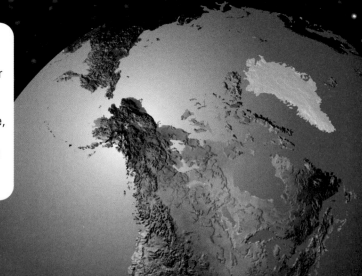